W9-DES-280

SUPERHERO ANIMALS

BATS

KENNY ABDO

Fly!
An Imprint of Abdo Zoom
abdobooks.com

abdobooks.com

Published by Abdo Zoom, a division of ABDO, P.O. Box 398166, Minneapolis, Minnesota 55439.

Printed in the United States of America, North Mankato, Minnesota.
102019
012020

Photo Credits: Alamy, Everette Collection, iStock, Shutterstock
Production Contributors: Kenny Abdo, Jennie Forsberg, Grace Hansen
Design Contributors: Dorothy Toth, Neil Klinepier

Library of Congress Control Number: 2019941316

Publisher's Cataloging-in-Publication Data

Names: Abdo, Kenny, author.
Title: Bats / by Kenny Abdo
Description: Minneapolis, Minnesota : Abdo Zoom, 2020 | Series: Superhero animals | Includes online resources and index.
Identifiers: ISBN 9781532129483 (lib. bdg.) | ISBN 9781098220464 (ebook) | ISBN 9781098220952 (Read-to-Me ebook)
Subjects: LCSH: Bats--Juvenile literature. | Bats--Behavior--Juvenile literature. | Nocturnal animals--Juvenile literature. | Sensory biology--Juvenile literature. | Animals--Flight--Juvenile literature. | Zoology--Juvenile literature.
Classification: DDC 599.45--dc23

TABLE OF CONTENTS

BATS

Guarding over Gotham City, Batman is the secret hero who lurks at night to ensure **justice** at all costs.

Bats are flying **mammals** found all around the world. There are more than 1,000 different **species**.

ORIGIN STORY

Batman flew into action in *Detective Comics* issue 27 published in 1939. Comic book writer Bill Finger and artist Bob Kane brought the hero to life.

Kane and Finger came up with the character by combining many influences, like Zorro, Sherlock Holmes, and The Shadow. Some of Batman's story was **autobiographical** to Kane.

POWERS & ABILITIES

Bats can be found in many different habitats. They live in forests, deserts, and jungles. They are even found in cities, like Gotham in the comics!

Bats are nocturnal. This means they are often active at night.

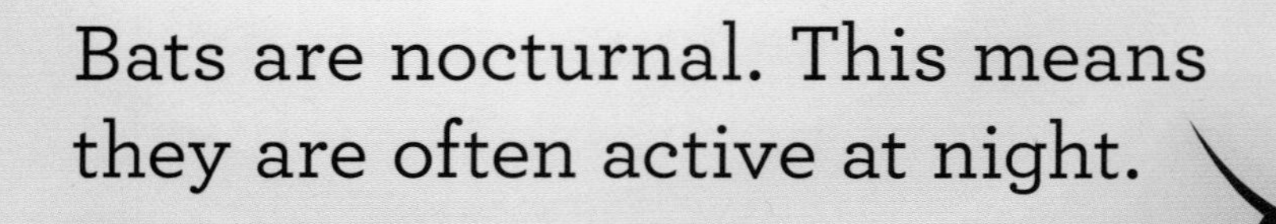

Bats see in the dark by using **echolocation**. Bats make noises that bounce back off of objects. They can tell how far something is by how quickly the sound returns to them.

Bats are **mammals** like humans. They are the only mammals capable of **sustained** flight. Some bats can reach flying speeds of 99 mph (160 kh)!

Bats help humans. They eat insects that ruin crops. Bats drop seeds that **pollinate** trees, flowers, and fruit plants. Their droppings can even be used as **fertilizer**!

IN ACTION

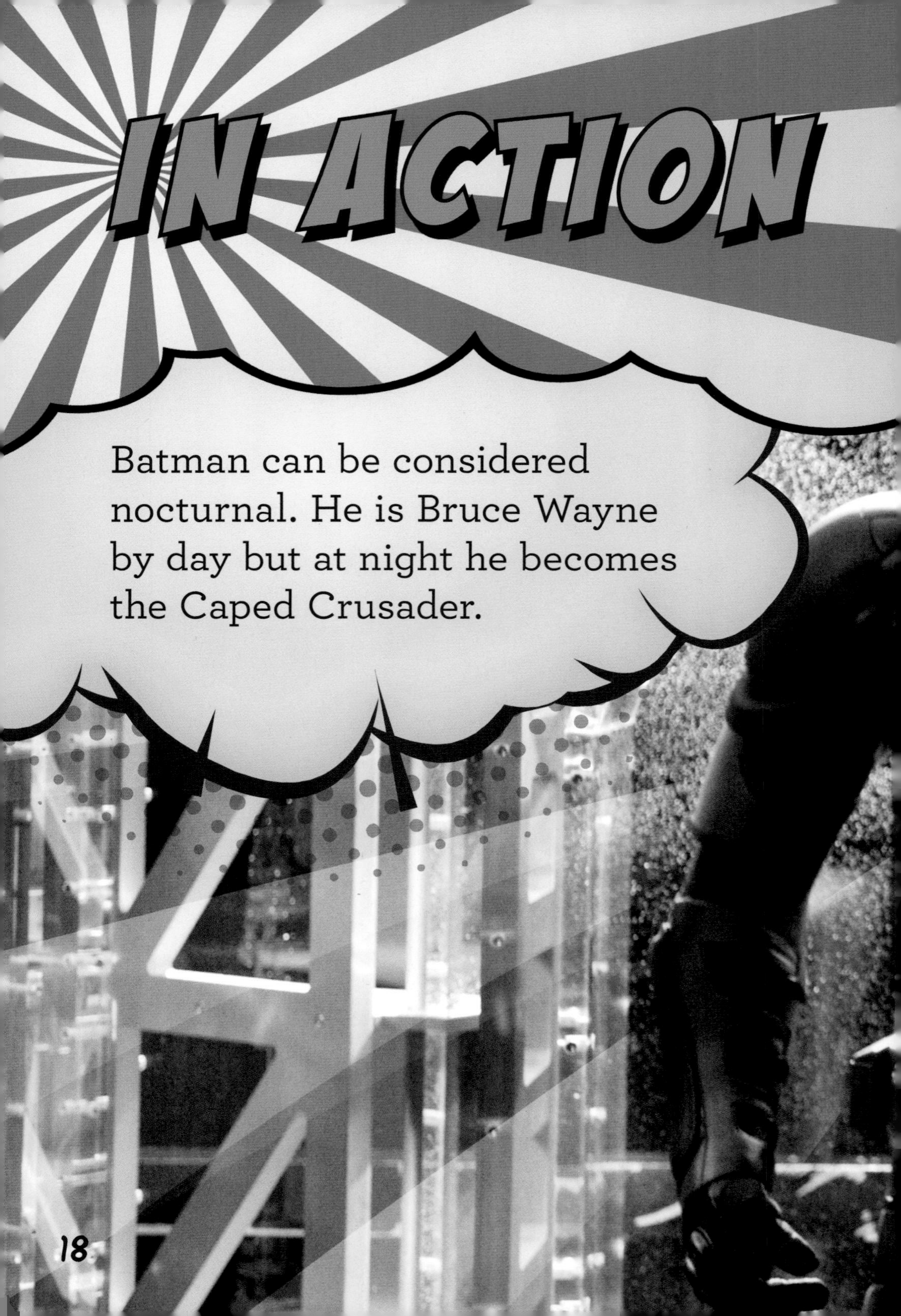

Batman can be considered nocturnal. He is Bruce Wayne by day but at night he becomes the Caped Crusader.

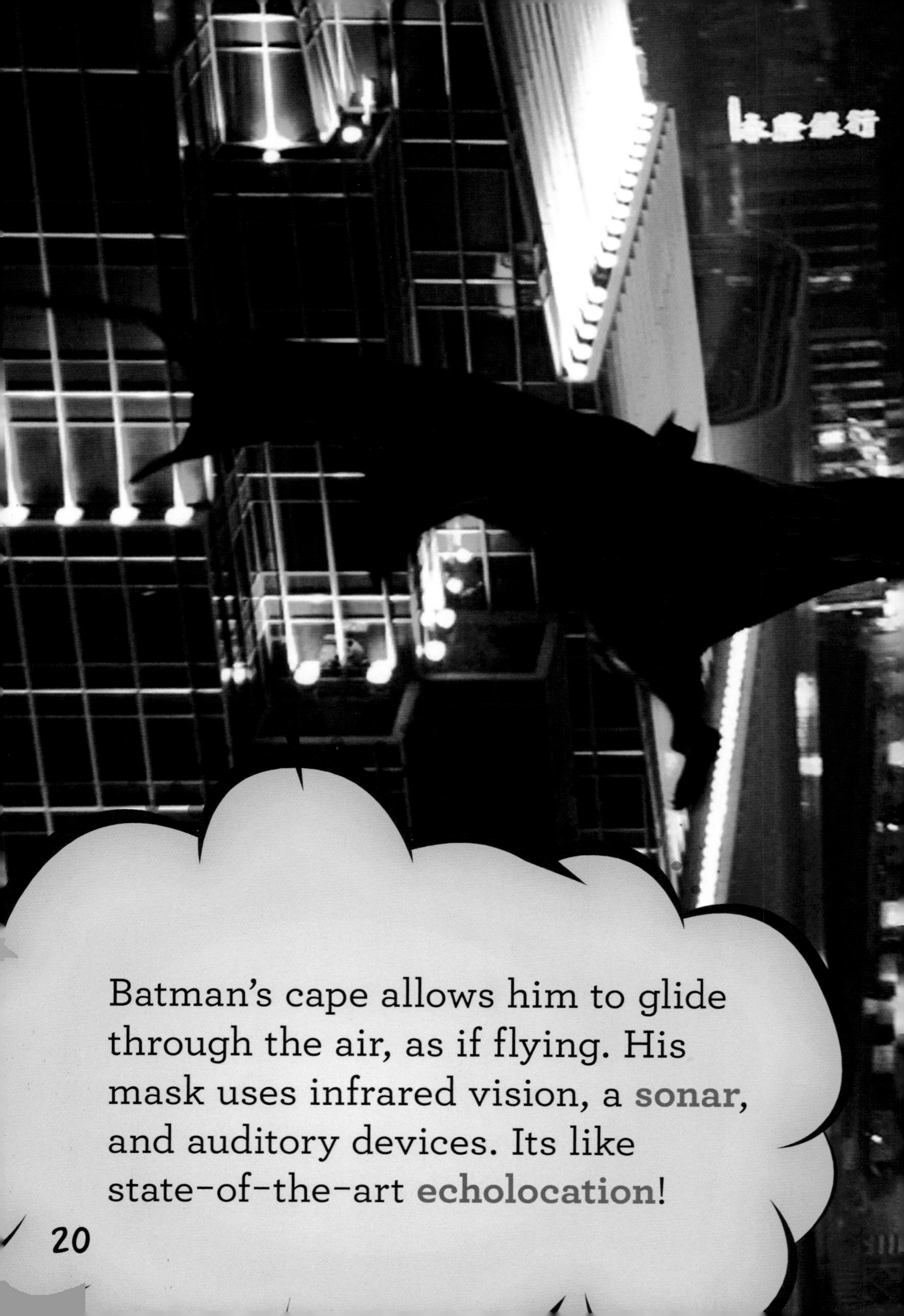

Batman's cape allows him to glide through the air, as if flying. His mask uses infrared vision, a **sonar**, and auditory devices. Its like state-of-the-art **echolocation**!

Above all else, Batman's only purpose is to help people. His unstoppable will to do good is what makes him not just a hero, but the Dark Knight!

GLOSSARY

autobiographical – marked by one's own experience or life history.

echolocation – the process by which a bat locates objects by emitting sounds and hearing them echoed.

fertilizer – something used to help plants grow.

justice – the upholding of what is fair.

mammal – a member of a group of living beings. Mammals make milk to feed their babies and usually have hair or fur on their skin.

pollinate – to carry pollen from one part of a flower or plant to another.

sonar – a way to find objects by sending and reflecting sound waves.

species – living things that are very much alike.

sustained – to maintain for a long period of time.

ONLINE RESOURCES

To learn more about bats, please visit **abdobooklinks.com** or scan this QR code. These links are routinely monitored and updated to provide the most current information available.

INDEX